YOUR KNOWLEDGE HAS VALUE

- We will publish your bachelor's and
  master's thesis, essays and papers

- Your own eBook and book -
  sold worldwide in all relevant shops

- Earn money with each sale

Upload your text at www.GRIN.com
and publish for free

Bibliographic information published by the German National Library:

The German National Library lists this publication in the National Bibliography; detailed bibliographic data are available on the Internet at http://dnb.dnb.de .

Imprint:

Copyright © 2017 GRIN Verlag, Open Publishing GmbH
Print and binding: Books on Demand GmbH, Norderstedt Germany
ISBN: 9783668479456

This book at GRIN:

http://www.grin.com/en/e-book/369802/phytochemical-analysis-and-cholesterol-lowering-efficiency-of-averrhoa

Prem Jose Vazhacharickal, Jiby John Mathew, Sajeshkumar N. K., Ratheesh Mohanan, Ellamplackill Surya Santhosh

Phytochemical analysis and cholesterol lowering efficiency of Averrhoa carambola Linn (star fruit).

An overview

GRIN Publishing

Phytochemical analysis and cholesterol lowering efficiency of *Averrhoa carambola* Linn. (star fruit): an overview

Prem Jose Vazhacharickal, Jiby John Mathew, Sajeshkumar N.K, Ratheesh Mohanan and Ellamplackill Surya Santhosh

ACKNOWLEDGEMENTS

Firstly we thank **God Almighty** whose blessing were always with us and helped us to complete this project work successfully.

We wish to thank our beloved Manager **Rev. Fr. Dr. George Njarakunnel,** Respected Principal **Dr. Joseph V.J,** Vice Principal **Fr. Joseph Allencheril,** Bursar **Shaji Augustine** and the Management for providing all the necessary facilities in carrying out the study. We express our sincere thanks to **Mr. Binoy A Mulanthra** (lab in charge, Department of Biotechnology) for the support. This research work will not be possible with the co-operation of many farmers.

Lastly, we extend our indebt thanks to patents, friends, and well wishers for their love and support.

Prem Jose Vazhacharickal*, Jiby John Mathew, Sajeshkumar N.K, Ratheesh Mohanan and Ellamplackill Surya Santhosh

*Address for correspondence
Assistant Professor
Department of Biotechnology
Mar Augusthinose College
Ramapuram-686576
Kerala, India

Table of contents

Table of figures

Table of tables

List of abbreviations

%	: Percentage
°C	: Degree celsius
µl	: Microlitre
Ca	: Calcium
CVD	: Cardio vascular disease
Fe	: Iron
H_2SO_4	: Sulphuric acid
HCl	: Hydrochloric acid
HDL	: High density lipoprotein
K	: Potassium
LDL	: Low density lipoprotein
N	: Nitrogen
Na	: Sodium
NaOH	: Sodium hydroxide
NH3	: Ammonia
P	: Phosphorous
SE	: Standard error

Phytochemical analysis and cholesterol lowering efficiency of *Averrhoa carambola* Linn. (star fruit): an overview

Prem Jose Vazhacharickal[1]*, Jiby John Mathew[1], Sajeshkumar N.K[1] , Ratheesh Mohanan[2] and Ellamplackill Surya Santhosh[1]

[1]Department of Biotechnology, Mar Augusthinose College, Ramapuram, Kerala, India-686576

[2]Department of Biochemistry, St. Thomas College, Pala, Kerala, India-686574

Abstract

Aqueous extract of the fruit pulp of *Averrhoa carambola* (star fruit) were evaluated for cholesterol lowering effect, in vitro, against various fatty food materials. Peoples are consuming food items made out of chicken, beef, mutton, egg and fish which contains large amount of fat. This study aims to analyze the effect of *Averrhoa carambola* in reducing the cholesterol level in this fat compound using water extract of the pulp. For this fatty food samples like egg yolk, pork fat, chicken fat , ghee and cod liver oil were treated with the extract and cholesterol level was estimated by Zak's method for a period of time. Phytochemical constituents present in water extract of *Averrhoa carambola* Linn. pulp includes alkaloids, saponins, steroids, phlobatannins, carbohydrate, terpenoids, phenols, coumarins, flavonoids and leucoanthocyanins. The in vitro cholesterol lowering effect of *Averrhoa carambola* pulp extract shows a positive result on reducing chicken fat, ghee and egg yolk. But in case of cod liver oil no beneficial change was observed.

Keywords: Cholesterol; Zak's method; Hypercholesterolemia, Star fruit, Emodins, Coumarins.

1. Introduction

Plants are the gifts of nature to human beings for basic preventive and curative Herbal medicine is the oldest form of healthcare known to mankind. Plants have always been an exemplary source of drugs and many of the currently available drugs have been derived directly or indirectly from them (Kamboj, 2000). Herbal medicines have often retained popularity for historical and cultural ingredients and are used primarily for treating mild and chronic ailments. India has an ancient heritage of traditional medicines; Materia Medica of India provides lots of information on the folklore practices and traditional aspects of therapeutically important natural products (Kirtikar and Basu, 2005). Indian Materia Medica includes about 2000 drugs of natural origin almost all of which are derived from different traditional systems and folklore practices. Out of these drugs derived from traditional system, 400 are of mineral and animal origin while the rest are of the vegetable origin. Natural products and especially those derived from higher plants have historically played a pivotal role in the discovery of new pharmaceuticals (Mukherjee, 2002). Undoubtedly, the plant kingdom still holds many species of plants containing substances of medicinal value which have yet to be discovered.

Star fruit or carambola (*Averrhoa carambola* Linn.) is a tropical fruit native to South East Asia, where it enjoys great popularity. Carambola fruit have an overall poor nutritional profile, except for their high vitamin C content which credits them with several wonderful health benefits. Vitamin C is a powerful natural antioxidant, anti-inflammatory and antimicrobial, with immune boosting properties and an anti-aging action. With good dietary fibre content, carambola helps lower blood cholesterol levels and contributes to colon health.

Research has shown that carambola has hypo-lipidemic effect and it prevents the development of non-alcoholic fatty liver disease. It was observed that treatment with carambola lowered the level of total cholesterol and triglyceride. Such effect was due to high content of dietary fibre. High dietary fibre levels present in carambola increased the excretion of total fat, cholesterol and bile acids. In addition to this, carambola contains phytosterols that help in lowering high cholesterol levels.

Cholesterol is a steroid lipid (fat) found in the blood of all animals and is necessary for proper functioning of our cell membranes and production of hormones. While there can be negative health benefits associated with low cholesterol, cholesterol deficiency is rare. Our bodies already manufacture all the cholesterol we need, so it is not necessary to consume more. Excessive consumption of cholesterol has been shown to increase the risk of heart disease and stroke. Cholesterol is only found in animal food products, and thus, vegans are likely to have lower cholesterol than non-vegans. High cholesterol foods include eggs, liver, fish, fast foods, butter, shellfish, shrimp, bacon, sausages, red meat, cheese, and pastries.

In Kerala most of the peoples are eating meat of different animals and birds, egg, coconut oils and fish etc that contain high amount saturated fat. This may the increase the blood cholesterol level and may lead to cardio vascular disease (CVD). Now a day's many people are taking star fruit to lower the cholesterol level. This study analysis the effect of   *Averrhoa carambola* fruit in cholesterol reduction.

1.1 Objectives

The objectives of this study to evaluate the phytochemical properties of methanol and water extract of *Averrhoa carambola* fruit extracts and its cholesterol lowering effect on various fatty food materials.

2. Review of literature

Herbal medicine is the oldest form of healthcare known to mankind. Plants have always been an exemplary source of drugs and many of the currently available drugs have been derived directly or indirectly from them (Kamboj, 2000). Herbal medicines have often retained popularity for historical and cultural ingredients and are used primarily for treating mild and chronic ailments. India has an ancient heritage of traditional medicines; 'Materia Medica' of India provides lots of information on the folklore practices and traditional aspects of therapeutically important natural products (Kirtikar and Basu, 2005). Indian 'Materia Medica' includes about 2,000 drugs of natural origin almost all of which are derived from different traditional systems and folklore practices. Out of these drugs derived from traditional system, 400 are of mineral and animal origin while the rest are of the vegetable origin (Mukherjee, 2002). Natural products and especially those derived from higher plants have historically played a pivotal role in the discovery of new pharmaceuticals. (Mukherjee, 2002).

Herbs are staging a comeback and herbal resistance is happening all over the world (Joy et al., 2001). In the Western world, as the people are becoming aware of the potency and side effects of synthetic drugs, there is an interesting interest in the natural product remedies with a basic approach towards the nature (Gheewala et al., 2012). In recent years, the use of herbal medicines worldwide has provided an excellent opportunity to India to look for therapeutic lead compounds from an ancient system of therapy; Ayurveda, which can be utilized for development of new drug. Over 50% of all modern drugs are of natural product origin and they play an important role in drug development programs of the pharmaceutical industry. (Mukerjee, 2002).

Plants have always been a significant source of natural products having therapeutic potential. Fruits, vegetables, nuts and grains are accounted to have many important biological effects including antioxidant, antitumor, anti-mutagenic and antimicrobial properties (Guanghou and Lai, 2002). Plant derived substances have recently become of great interest owing to their versatile applications. Medicinal plants are the richest bio- resource of drugs of traditional systems of medicine, modern medicines, nutraceuticals, food supplements, folk medicines, pharmaceutical intermediates and chemical entities for synthetic drugs (Ncube et al., 2008)

Cholesterol is a compound involved in the essential functions of the human body. Cholesterol travels through the blood via the arteries, which is why a high cholesterol diet can lead to clogged arteries that prevent proper blood flow to your heart, leading to heart disease and even diabetes. A plant based diet is essential to reduce cholesterol. Some plants used are Lutein-Rich Spinach it helps reduce cholesterol by preventing clogged arteries. Walnuts are a fantastic source of omega 3 fats. These fats lower cholesterol directly by reducing inflammation and raising good cholesterol. Beans are the highest fibre food. Fibre has been shown to improve heart health and directly lower high cholesterol in a short amount of time. Other cholesterol lowering items are Avocado, Natural Soy foods, Berries, Cacao and Garlic.

Blood cholesterol is a fatty substance produced naturally by your liver and found in your blood. Blood cholesterol is used for many different things in your body, but it can become a problem when there is too much of it in your blood. There are mainly two types, that is, low density lipoprotein (LDL) also known as bad cholesterol because it can add to the build-up of plaque in your arteries and increases your risk of getting coronary heart diseases. And other type is high-density lipoprotein (HDL) also known as good cholesterol because it can help to protect you against coronary heart disease. In addition, your blood also contains a type of fat called triglycerides, which are stored in your body's fat deposits (Victoria, 2016)

There are many causes of high cholesterol such as low intake of foods containing healthy, protective fats-this increases your intake of polyunsaturated and monounsaturated fats, which tend to increase the HDL cholesterol in your blood, high intake of foods containing saturated fats and trans-fats; such as fatty meats, full-fat dairy products, butter, coconut oil, palm oil and most deep-fried takeaway foods and commercially baked products, such as pies, biscuits, buns and pastries. Foods with high trans-fats include most deep-fried takeaway foods and commercially baked products. High cholesterol level can lead to increased risk of coronary heart diseases, heart attack and stroke. As the lipids block the arteries which stop the supply of blood (Victoria, 2016).

Star fruit also known as *Averrhoa carambola* is a rich source of vitamin C. It is a tropical fruit with low nutritional value except high abundance of vitamin C content which credits them with several wonderful health benefits. With high dietary fibre content, carambola helps lower blood cholesterol levels and contributes to colon health. Carambola contains 2.8 g of dietary fibre/100 g of fruit. Dietary fibre passes undigested through the digestive tract and binds to lipids (fat) form food, preventing their absorption at the intestinal level. This helps lower cholesterol levels, contributing to cardiovascular health (Lixandru, 2015).

Underutilized tropical fruits provide limitless opportunities for screening of novel drugs. Five-lobbed fleshy, yellow-greenish, edible fruits of *Averrhoa carambola* Linn. (star fruit) of Oxalidaceae is native of South-East Asia and cultivated in some parts of India. The fruits are good source of antioxidants and used traditionally in mouth ulcers, toothache, nausea, diarrhoea, ascites etc. (Shui and Leong, 2006). Edible fruits of *Zizyphus mauritiana* Lam. (jujube or ber) of Rhamnaceae, an evergreen perennial tree native to India is used for various ethno-medical uses in asthma, burning, boils, nausea, vomiting and piles.

Carambolas have been cultivated in South East Asia (Malaysia, India, Sri Lanka) for centuries and trees were introduced in Florida over 100 years ago. Fruit from the first introductions into Florida were tart. More recently, seeds and vegetative material from Thailand, Taiwan, and Malaysia have been introduced and sweet cultivars have been selected (Dasgupta et al., 2013; Gheewala et al., 2012)

Description for star fruit are as follows: Tree: The carambola tree is small to medium in height (22 to 33 ft; 7 to 10 m), spreading (20 to 25 ft in diameter; 6 to 7.6 m), evergreen, and single or multi-trunked. Trees grow rapidly in locations protected from strong winds. The mid-canopy area (3 to 7 ft height; 0.9 to 2.1 m) is the major fruit producing area of mature trees.

Leaves: Carambolas have compound leaves 6 to 12 inches (15 to 30 cm) long that are arranged alternately on branches. Each leaf has 5 to 11 green leaflets 0.5 to 3.5 inches long (1.5 to 9 cm long) and 0.4 to 1.8 inches (1 to 4.5 cm) wide.

Flowers: Carambola flowers are born on panicles, twigs, small diameter branches, and occasionally on larger wood. Flowers are perfect, small (3/8 inch or 1 cm in diameter), pink to lavender in colour, and have 5 petals and sepals. Depending upon the cultivar, carambola flowers have either long or short styles (Hasim, 2004).

Fruit: The fruit is a fleshy 4 to 5 celled berry with a waxy surface. Fruit are 2 to 6 inches (5-15 cm) in length, with 5 (rarely 4-8) prominent longitudinal ribs; star-shaped in cross section. The fruit skin is thin, light to dark yellow, and smooth, with a waxy cuticle.

Averrhoa carambola is fully packed with vital nutrients. It is a very good source of natural antioxidants like L-ascorbic acid, (-) epicatechin and gallic acid in gallotannin forms (Guanghou, 2004). Consuming 100 g of this fruit can provide, 35.7 g calories, 0.38 g proteins, 9.38 g carbohydrates, 0.80-0.90 g dietary fibre, 0.8 g fat, 4.4-6.0 mg Ca (calcium), 0.321.65 mg Fe (iron), 15.5-21.0 mg P (phosphorus), 2.35 mg K (potassium), 0.003-0.552 mg of carotene, 4.37 mg tartaric acid, 9.6 mg oxalic acid, 2.2 mg α-ketoglutaric acid, 1.32 mg citric acid. Moreover, various amino acids like 0.03-0.038 mg of thiamine, 0.0190.03 mg of riboflavin, 0.294-0.38 mg of niacin, 3 mg of tryptophan, 2 mg of methionine and 26 mg of lysine are also present in 100 g of the fruit. The fruit is famous for very low fat, rich in vitamin B and vitamin C and also a source for potassium and fibre (Crane, 1994).

Averrhoa.Carambola have been cloned to different hybrids such as B2 (Maha 66), B10, B17 and B11. It yields 300 fruits per tree at age of 2 years until 9 years. Its fruiting season is mainly June and October to December. Star fruit has many uses such as it can be consumed as juice, used for production of jams and jellies etc, used to make wine and brandy, may lead to lowering of blood pressure, its consumption can be used to relief headache, vomiting, coughing and restlessness. It is also used to halt haemorrhages and to relieve bleeding haemorrhoids, useful as a treatment for fever, eczema, haemorrhages, haemorrhoids and diarrhoea. It also has its disadvantages, over consumption of star fruit can lead to intoxication such as severe low back pain radiating to both legs, physical examination revealed with positive bilateral straight leg raising test, increase blood urea N (nitrogen), Na (sodium), K, glucose, and it decreases blood pressure, pulse, reparatory rate and body temperature.

In the recent research conducted it was found that star fruit intoxication causes consciousness disturbance in patients with renal failure. It is caused by the neurotoxin named as caramboxin, a non-protein molecule causes the renal diseases. Such uremic patients are cured through continuous dialysis till it is cured. Hence hemo-dialysis is done at the first stage itself or else it can cause renal failure (Chang et al., 2002).

Phytochemicals are of two categories i.e., primary and secondary constituents. Primary constituents have chlorophyll, proteins sugar and amino acids. Secondary constituents contain terpenoids and alkaloids. Terpenoids exhibit various important pharmacological activities i.e., anti-inflammatory, anti-cancer, anti-malarial, inhibition of cholesterol synthesis, anti-viral and anti-bacterial activities (Gheewala et al., 2012). Terpenoids are very important in attracting useful the herbivorous insects and mites. Alkaloids are used as anaesthetic agents and are found in medicinal plants. (Gheewala et al., 2012).

In phytochemical studies anti-ulcer properties, presence of triterpenes, flavonoids and mucilage was observed (Verma and Singh, 2008). However most of the work done so far has not been followed up in such a way to clear scientific doubts and determine active principles and mechanism of action. In view of the nature of plant more research can be done to investigate the unexplored and unexploited potential of plant. Only such research would place *Averrhoa carambola* in its proper place in nutritional and medical sciences (Gheewala et al., 2012).

3. Hypothesis

The current research work is based on the following hypothesis

1) *Averrhoa carambola* extracts are rich in various phytochemical components.
2) These extract could lower cholesterol levels.

4. Materials and Methods

4.1 Study area

Kerala state covers an area of 38,863 km^2 with a population density of 859 per km^2 and spread across 14 districts. The climate is characterized by tropical wet and dry with average annual rainfall amounts to 2,817 ± 406 mm and mean annual temperature is 26.8°C (averages from 1871-2005; Krishnakumar et al., 2009). Maximum rainfall occurs from June to September mainly due to South West Monsoon and temperatures are highest in May and November.

4.2 Collection of plant material

The fruits of *Averrhoa carambola* were collected from Thalanadu, kottayam district of Kerala state, India during the month of July/ August in the year 2016 and identified using the taxonomic descriptors. The fresh ripen fruits were used.

4.3 Preparation of *Averrhoa carambola* fruit pulp extracts

For the preparation of extract, two different solvents; water and ethanol. The dried fruits of *Averrhoa carambola* were cut into very small pieces using a sterile knife. Ten grams of the dried fruits were transferred into separate bottles containing 50 ml of the solvents. The mixture was kept in a shaker at room temperature (30°C) at 170 rpm for 24 hours for continuous shaking. The extracts were collected in sterile tubes in a laminar air flow chamber and were centrifuged at 4900 rpm for 10 minutes to remove the debris. The supernatant containing the extracts were collected in a sterile tube and stored at 4°C for further use.

4.4 Phytochemical screening

Chemical presence or absence of certain important compounds in an extract is determined by colour reactions of the compounds with specific chemicals which act as dyes. This procedure is a simple preliminary prerequisite before going for detailed phytochemical investigation. Various tests have been conducted qualitatively to find out the presence or absence of bioactive compounds.

Chemical tests were carried out on the aqueous extract specimens using standard procedures to identify the constituents as described by Sofowara (1993), Trease and Evans (1989), Harborne (1973), Roy and Geetha (2013), Vazhacharickal et al. (2016), and Mathew et al. (2016).

4.4.1 Test for alkaloids

Two ml of plant extract was taken in a test tube and few drops of Hager's reagent were added. Yellow precipitate shows positive result for alkaloids.

4.4.2 Test for anthraquinones

Three ml of plant extract was taken in a test tube and three ml of benzene and five ml of ten percentage NH_3 (ammonia) were added. Formation of pink, violet or red coloration in ammonical layer detect the presence of anthraquinones.

4.4.3 Test for anthocyanins

Two ml of plant extract was taken in a test tube and two ml of 2N HCl (hydrochloric acid) and NH_3 were added. Formation of pinkish red to bluish violet coloration indicates the presence of anthocyanins.

4.4.4 Test for carbohydrate

Two ml of plant extract was taken in a test tube and ten ml of water, two drops of twenty percentage ethanolic α naphthol and two ml of conc.H_2SO_4 (sulphuric acid) were added. Formation of reddish violet ring at the junction shows the presence of carbohydrates.

4.4.5 Test for coumarins

Two ml of extract was taken in a test tube and three ml of ten percentage NaOH (sodium hydroxide) was added. Formation of yellow colour gives positive result to coumarins.

4.4.6 Test for emodins

Two ml of plant extract was taken in a test tube and two ml of NH_4OH (ammonium hydroxide) and three ml of benzene were added. Formation of red colour indicates the presence of emodins.

4.4.7 Test for flavonoids

Five ml of dilute ammonia solution were added to a portion of the plant extract followed by addition of concentrated H_2SO_4. A yellow colouration observed in each extract indicated the presence of flavonoids. The yellow colouration disappeared on standing.

4.4.8. Test for glycosides

Two ml of plant extract was taken in a test tube and two ml of chloroform and two ml of acetic acid were added. Formation of violet to blue to green coloration shows the presence of glycosides.

4.4.9 Test for leucoanthocyanins

Five ml of isoamyl alcohol taken in a test tube and five ml of plant extract was added. Turn organic layer into red detects the presence of leucoanthocyanins.

4.4.10 Test for phlobatannins

Deposition of a red precipitate when an extract of each plant sample was boiled with one percentage aqueous HCl was taken as evidence for the presence of phlobatannins.

4.4.11 Test for proteins

One ml of plant extract was mixed with one ml of conc.H_2SO_4 in a test tube. Formation of white precipitate indicate the presence of proteins

4.4.12 Test for phenols

Few ml of the plant extract was taken in attest tube and few ml of lead acetate was added to it. Formation of white precipitate detects the presence of phenols.

4.4.13 Test for saponins

Ten ml of the extract was mixed with five ml of distilled water and shaken vigorously for a stable persistent froth. The frothing was mixed with three drops of olive oil and shaken vigorously, then observed for the formation of emulsion.

4.4.14 Test for steroids

Two ml of extract was taken in a test tube and two ml chloroform and two ml of conc.H_2SO_4 was added. Formation of reddish brown ring at the junction shows the presence of steroids.

4.4.15 Test for terpenoids

Five ml of each extract was mixed in two ml of chloroform, and 3 ml concentrated H_2SO_4 was carefully added to form a layer. A reddish brown colouration of the inter face was formed to show positive results for the presence of terpenoids.

4.5 Preparation of cholesterol samples

One gram of the sample was dissolved in one ml of chloroform and stored in brown bottle for further use (Varley, 2004).

4.6 Treatment

200 µl of extract was added to each of the sample prepared and mixed well. These are used for the periodic (24 hour interval) determination of cholesterol.

4.7 Estimation of cholesterol

The amount of cholesterol in each sample was estimated by Zak's method before and after treatment (Varley, 2004).

4.8 Statistical analysis

The survey results were analyzed and descriptive statistics were done using SPSS 12.0 (SPSS Inc., an IBM Company, Chicago, USA) and graphs were generated using Sigma Plot 7 (Systat Software Inc., Chicago, USA).

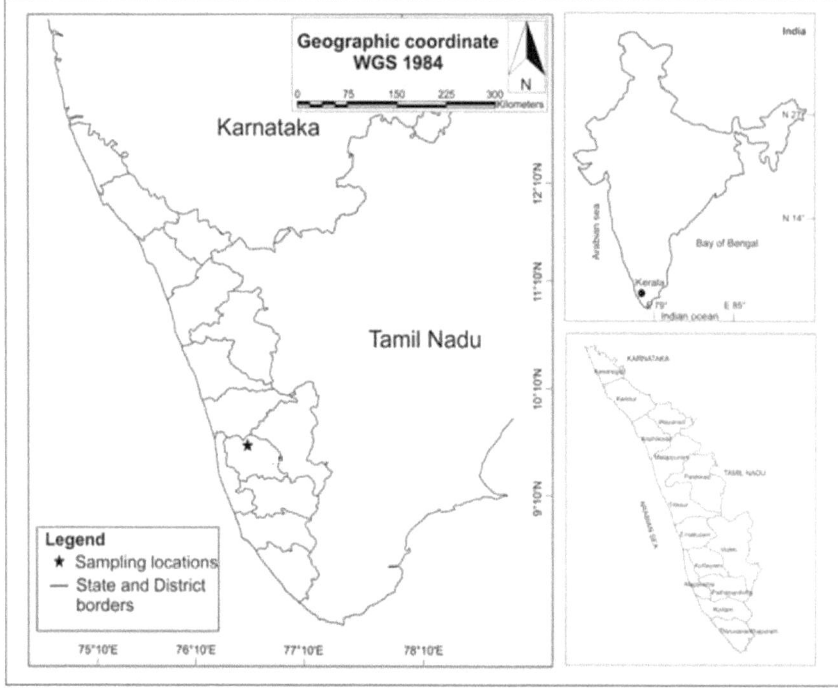

Figure 1. Map of Kerala showing the soil sample collection point. Authors own work.

Figure 2. Details of *Averrhoa carambola* Linn (star fruit). Photo courtesy: Wikipedia. https://en.wikipedia.org/wiki/Averrhoa_carambola.

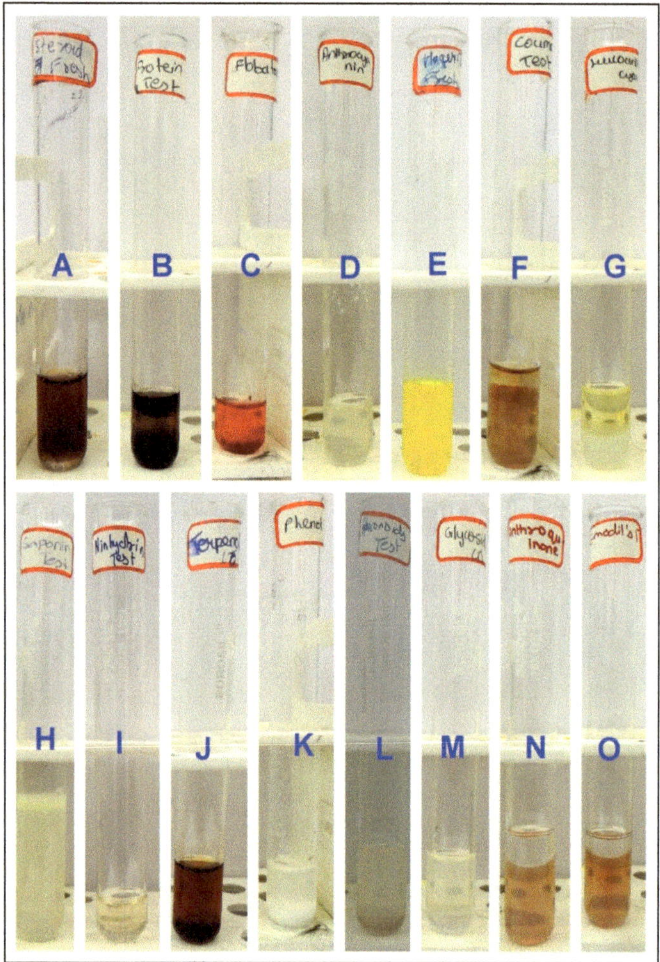

Figure 3. Details of phytochemical analysis of *Averrhoa carambola* fruit (methanol extract); A. Steroids, B. Proteins, C. Phlobatannins, D. Anthocyanines, E.Carbohydrates, F. Coumarins, G. Leucoanthocyanines, H. Saponins, I.Ninhydrin test, J. Terpenoids, K. Phenols, L.Flavonoids, M. Glycosidedes, N. Anthroquinine test, O. Emodines. Authors own images.

Table 1. Preliminary phytochemical analysis of *Averrhoa carambola* fruit extract.

Sl. NO.	Phytoconstituent	Fresh fruit extract of *Averrhoa carambola*
1.	Alkaloids	+
2.	Anthocyanins	-
3.	Anthraquinones	+
4.	Carbohydrates	+
5.	Coumarins	+
6.	Emodins	+
7.	Flavonoids	+
8.	Glycosides	-
9.	Leucoanthocyanins	-
10.	Phenol	+
11.	Phlobatannins	+
12.	Proteins	-
13.	Saponins	+
14.	Steroids	+
15.	Terpenoids	+

+ indicates presence of phytochemicals

- indicates absence of phytochemicals

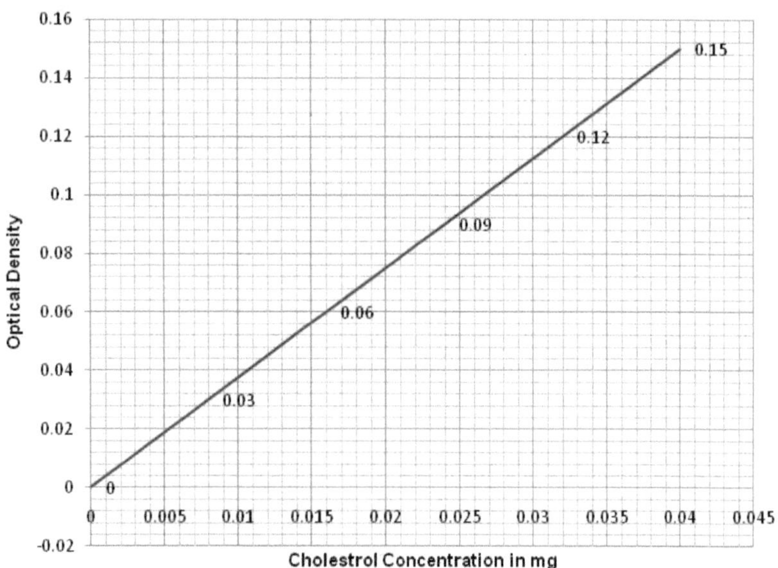

Figure 4. Standard graph for cholesterol estimation by Zak's method.

Table 2. Cholesterol estimation at different time intervals after treatment (n=3; values in mg/g sample).

Day	Substrate				
	Egg yolk	Pork Fat	Chicken Fat	Ghee	Cod liver oil
1	0.024 ± 0.01	0.016 ± 0.01	0.027 ± 0.03	0.021 ± 0.02	0.012 ± 0.03
2	0.024 ± 0.01	0.016 ± 0.01	0.024 ± 0.02	0.012 ± 0.01	0.012 ± 0.01
3	0.024 ± 0.01	0.012 ± 0.02	0.021 ± 0.01	0.009 ± 0.01	0.012 ± 0.01
4	0.021 ± 0.01	0.009 ± 0.01	0.021 ± 0.01	0.009 ± 0.01	0.012 ± 0.01
5	0.021 ± 0.01	0.009 ± 0.00	0.019 ± 0.01	0.009 ± 0.01	0.012 ± 0.00
6	0.019 ± 0.02	0.009 ± 0.01	0.019 ± 0.01	0.008 ± 0.01	0.012 ± 0.00
7	0.016 ± 0.01	0.009 ± 0.00	0.016 ± 0.01	0.008 ± 0.01	0.012 ± 0.00
8	0.012 ± 0.02	0.005 ± 0.01	0.012 ± 0.01	0.008 ± 0.01	0.012 ± 0.01
9	0.012 ± 0.00	0.005 ± 0.01	0.012 ± 0.00	0.008 ± 0.01	0.012 ± 0.01
10	0.008 ± 0.01	0.001 ± 0.01	0.009 ± 0.01	0.005 ± 0.01	0.012 ± 0.01
11	0.005 ± 0.01	0.001 ± 0.00	0.001 ± 0.00	0.005 ± 0.01	0.012 ± 0.01
12	0.002 ± 0.01	0.001 ± 0.00	0.001 ± 0.00	0.005 ± 0.00	0.012 ± 0.00

Numbers represent means ± one standard error (SE) of mean.

5. Results and discussion

The preliminary phytochemical screening of the crude fresh extract of *Averrhoa carambola* revealed the presence of various phytoconstituents including alkaloids, anthraquinones, carbohydrates, coumarins, emodins, flavonoids, phenol, phlobatannins, saponins, steroids and terpenoids in them.

The in vitro cholesterol lowering analysis shows that the fruit extract has an ability to reduce the total cholesterol present in the fatty food samples. In ghee, the amount of cholesterol was reduced to 1/3 after 12 days of treatment. chicken fat, egg yolk and pork fat also shows considerable reduction in cholesterol level. But in case of cod liver oil no considerable change were found.

Studies revealed that Carambola contains phytosterols that help in lowering high cholesterol levels. These sterols are similar to cholesterol and it occurs in plants only. Many studies have claimed that phytosterols reduce the level of LDL cholesterol (unhealthy) and thus, reduces the overall risk of cardiovascular diseases. Another research showed that health promoting compounds present in carambola fruit such as phytochemicals, flavonoids, phytosterols, saponins, ascorbic acid and dietary fibre either increase the use of fat as a source of energy or increase excretion of excess cholesterol out of the body or thus, reduces its storage in the body in the form of cholesterol. Thus, carambola fruit can be made an important part of cholesterol lowering diet. Any of these compounds may have in vitro cholesterol degrading property that should be subjected to further investigation.

Averrhoa carambola is a plant with diverse potentials. The wide spectrum of phytochemicals bottled up in various parts of the plant makes it ideal for nutritional and medicinal uses which could help in salvaging the mankind from the present burden of diseases. Numerous phytochemical and pharmacological studies have been conducted on different parts of *Averrhoa carambola*. However most of the work done so far has not been followed up in such a way to clear scientific doubts and determine active principles and mechanism of action. In view of the nature of plant more research can be done to investigate the unexplored and unexploited potential of the plant. Only such research would place *Averrhoa carambola* in its proper place in nutritional and medical sciences.

6. Conclusions

The study was conducted to analyse the phytochemical characteristics of *Averrhoa carambola* and evaluates its cholesterol lowering activity. The plant parts have benefited mankind in various ways, including treatment of many diseases. Most of the parts like leaves, bark, flowers, fruits, seeds, roots or the whole plant are used as alternative medicine to treat a variety of diseases. The fruits and leaves are the most valuable parts and have often used as a curative agent in several pharmacopeias.

Phytochemicals are biologically active, naturally occurring chemical compounds found in plants, which provide health benefits. Phytochemical tests were carried out on the aqueous extract and standard procedures used to identify the constituents. Various phytochemical compounds like alkaloids, flavonoids, saponins, glycosides, emodins, proteins, carbohydrate, terpenoids, tannins, coumarins and phenols were found in the fruit extracts of the plant.

Five common fatty food materials like egg yolk, pork fat, chicken fat, ghee and cod liver oil was mixed with the fruit extract of *Averrhoa carambola* and the cholesterol level in each was evaluated in vitro, after and before treatment. The in vitro analysis shows that cholesterol level of fatty food materials treated with the extract show reducing day by day except cod liver oil. The change in cholesterol level was not evident in the early days but significant reduction was observed after a week.

References

Chang, C. T., Chen, Y. C., Fang, J. T., & Huang, C. C. (2002). Star fruit (*Averrhoa carambola*) intoxication: an important cause of consciousness disturbance in patients with renal failure. *Renal Failure*, 24(3), 379-382.

Chang, C. T., Chen, Y. C., Fang, J. T., & Huang, C. C. (2002). Star fruit (*Averrhoa carambola*) intoxication: an important cause of consciousness disturbance in patients with renal failure. *Renal failure*, 24(3), 379-382.

Crane, J. H. (1994).The Carambola, University of Forida, Fact sheet HS-12, april 1994.

Dasgupta, P., Chakraborty, P., & Bala, N. N. (2013). *Averrhoa carambola*: an updated review. *International Journal of Pharma Research & Review*, 2(7), 54-63.

Gheewala, P., Kalaria, P., Chakraborty, M., & Kamath, J. V. (2012). Phytochemical and Pharmacological profile of *Averrhoa carambola* Linn: An overview. *International Research Journal of Pharmacy*, 3(1), 88-92.

Harborne, J. B. (1973). Phenolic compounds. In *Phytochemical methods* (pp. 33-88). Springer Netherlands.

Joy, P. P., Mathew, S. & Skaria, B.P. (2001). Medicinal Plants. In: Bose, T.K., Kabir, J., Das, P. & Joy, P.P. (editors). Tropical horticulture. Vol 2, Naya Prokash, Calcutta, India.

Kamboj, V. P. (2000). Herbal medicine. *Current Science*, 78(1), 35-39.

Kirtikar, K. R. & Basu, B. D. (2005). Indian medicinal Plant. International Book Distributors, Dehradun, India.

Krishnakumar, K. N., Rao, G. P., & Gopakumar, C. S. (2009). Rainfall trends in twentieth century over Kerala, India. *Atmospheric Environment*, 43(11), 1940-1944.

Krishnakumar, T., Jayaprakash, R., Pinna, N., Singh, V. N., Mehta, B. R., & Phani, A. R. (2009). Microwave-assisted synthesis and characterization of flower shaped zinc oxide nanostructures. *Materials Letters*, 63(2), 242-245.

Lixandru, M. (2015). Properties and benefits of carambola, star fruit for high blood cholesterol. Available from: https://www.natureword.com/tag/star-fruit-for-high-blood-cholesterol/

Mathew, J. J., Vazhacharickal, P. J., Sajeshkumar N. K. & Joy, J. K. (2016). Phytochemical analysis and In vitro hemostatic activity of *Mimosa pudica*,

Hemigraphis colorata and *Chromolaena odorata* leaf extracts; *CIBTech Journal of Pharmaceutical Sciences*, 5(3), 16-34.

Morton, J. F. (2004). Fruits of warm climate. Flair Books, USA.

Mukherjee, P. K. (2002). Quality control of herbal drugs: an approach to evaluation of botanicals. Business Horizons Publication, New Delhi, India.

Ncube, N. S., Afolayan, A. J., & Okoh, A. I. (2008). Assessment techniques of antimicrobial properties of natural compounds of plant origin: current methods and future trends. *African Journal of Biotechnology*, 7(12), 1797-1806.

Roy, A., & Geetha, R. V. (2013). Preliminary Phytochemical screening of the Ethanolic extract of *Dioscorea villosa* Tubers and Estimation of Diosgenin by HPTLC Technique. *International Journal of Drug Development and Research*, 5(3), 377-381.

Saunders, E. W. C. (1997). Text book of Pharmacognosy. 14th ed. Harcourt Brace Asia, Singapore.

Sofowora, A. (1993). Recent trends in research into African medicinal plants. *Journal of Ethnopharmacology*, 38(2-3), 197-208.

Trease, G. E., & Evans, W. C. Pharmacognosy. 1989. *Bailliere Tindall, London*, 45-50.

Varley, H. (2004). Practical clinical Biochemistry, 4th edition, Heinemann Medical, UK.

Vazhacharickal, P. J., Mathew, J. J., Sajeshkumar N.K & Varghese, N. (2015). A comparative study on traditional herbal medicines of Mannans (tribal group in Idukki, Kerala) for the treatment of diabetes, Grin Publishing, Germany.

Verma, S., & Singh, S. P. (2008). Current and future status of herbal medicines. *Veterinary World*, 1(11), 347-350.